TOP-SECRET NATURE

WHY THE SUN SETS

MARIE ROGERS

PowerKiDS press

New York

Published in 2021 by The Rosen Publishing Group, Inc.
29 East 21st Street, New York, NY 10010

First Edition

Editor: Amanda Vink
Book Design: Rachel Rising

Portions of this work were originally authored by Violet Miller and published as *Why Does The Sun Set?* All new material in this edition authored by Marie Rogers.

Photo Credits: Cover, p. 1 Creative Travel Projects/Shutterstock.com; pp. 4,6,8,10,12,14,16,18,20,22 (background) cluckva/Shutterstock.com; p. 5 Edmund O'Connor/Shuttestock.com; p. 7 Vladi333/Shutterstock.com; p. 9 C. Rene Ammundsen/Shutterstock.com; p. 11 Triff/Shutterstock.com; p. 13 voyata/Shutterstock.com; p.13 Victoria Lipov/Shutterstock.com; p. 15 VectorMine/Shutterstock.com; p. 17 biletskiy/Shutterstock.com; p. 19 Pakhnyushchy/Shutterstock.com; p. 21 LeAndr/Shutterstock.com; p. 22 vovan/Shutterstock.com.

Library of Congress Cataloging-in-Publication Data

Names: Rogers, Marie, 1990- author.
Title: Why the sun sets / Marie Rogers.
Description: New York : PowerKids Press, [2021] | Series: Top-secret nature | Includes index.
Identifiers: LCCN 2019050017 | ISBN 9781725317659 (paperback) | ISBN 9781725317673 (library binding) | ISBN 9781725317666 (6 pack) | ISBN 9781725317680 (ebook)
Subjects: LCSH: Earth (Planet)–Rotation–Juvenile literature.
Classification: LCC QB633 .R58 2021 | DDC 525/.35–dc23
LC record available at https://lccn.loc.gov/2019050017

Manufactured in the United States of America

CPSIA Compliance Information: Batch #CSPK20. For Further Information contact Rosen Publishing, New York, New York at 1-800-237-9932.

CONTENTS

Again and Again

During a day on Earth, the sun seems to move across the sky. The sun comes up in the morning in the east and sets at night in the west. Every day, the same thing happens again. But why?

Earth's Rotation

While it may look like it from the ground, the sun isn't really moving through the sky! It only seems like that because Earth rotates. When something rotates, it turns around a center point or line.

The Speed of Rotation

Human beings don't feel Earth moving. That's because everything on Earth, including humans, moves too. The **speed** of rotation depends on where you are on Earth. On the **equator**, Earth spins faster than it does at the **poles**.

NORTH AMERICA
HUDSON BAY
UNITED STATES
GULF OF MEXICO
SARGASSO SEA
NORTH ATLANTIC OCEAN
Columbus 1492-93
Christopher Columbus 1451-1506
BAHAMAS
CUBA
JAMAICA
WEST INDIES
CARIBBEAN SEA
CENTRAL AMERICA
BELIZE
COLOMBIA
VENEZUELA
GUYANA
SURINAME
ECUADOR
PERU
SOUTH AMERICA
BRAZIL
Tropic of Cancer
90°W.

Earth's Axis

Earth spins on its axis. An axis is the imaginary line around which something rotates. Earth's axis passes through the North Pole and the South Pole. Earth isn't straight up and down—its axis is **tilted**! That makes sunlight hit areas of Earth differently.

Different Sides

One rotation lasts 24 hours, or a day. Because Earth's round, not all places face the same way at the same time. It's daytime on the side of the world facing the sun. It's nighttime on the side of the world facing away from the sun.

Day Lengths

Places at the equator have just about the same amount of daylight all year long. At the poles of Earth, the amount of daylight changes at different times of the year. The sun doesn't set during summer at the North Pole!

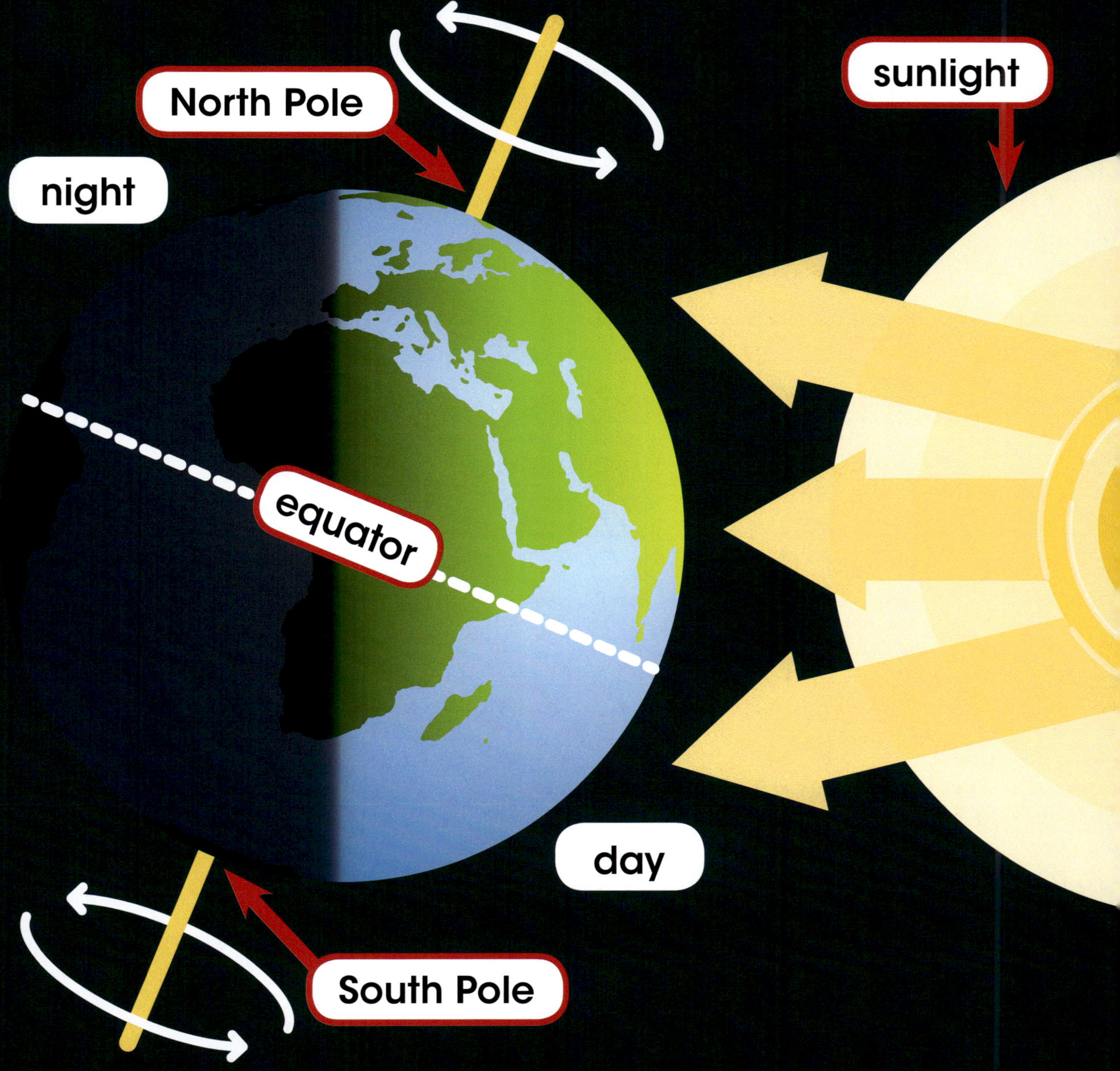
North Pole
night
sunlight
equator
day
South Pole

Sunrise

As the sun rises, the sky goes from dark to light. When the sky gets lighter, it's dawn. The sun appears and begins to rise. The sun gets higher for the first half of the day. It's midday at the sun's highest point.

Midday

The sun moves lower in the sky as Earth rotates away from it. In the summer, the sun rises higher in the sky than in the winter. That's because of Earth's tilted axis. Some areas face the sun more directly at different times of the year.

Sunset

As Earth continues to rotate, the sun begins to set. This is called dusk. Sunsets can be very beautiful. They can look red, yellow, and orange. Even after the sun has left the sky, sunlight colors the sky for a little while longer.

A New Day

After the sun has set, we begin to see the moon and stars. They give us light. They look like they move, too! That's because Earth spins all night. In the morning, the sun rises to greet us again.

GLOSSARY

equator: An imaginary circle around Earth that's an equal distance from the North Pole and the South Pole.

pole: Either end of the imaginary line on which Earth turns.

speed: The rate at which someone or something moves.

tilt: To move so as to slant or tip.

INDEX

WEBSITES

Due to the changing nature of Internet links, PowerKids Press has developed an online list of websites related to the subject of this book. This site is updated regularly. Please use this link to access the list: www.powerkidslinks.com/tsn/sunset